BEES
& BEEKEEPING EXPLAINED

GERARD BAKER

COUNTRYSIDE BOOKS
NEWBURY BERKSHIRE

First published 2010
© Gerard Baker 2010

All rights reserved. No reproduction
permitted without the prior permission
of the publisher:

COUNTRYSIDE BOOKS
3 Catherine Road
Newbury, Berkshire

To view our complete range of books,
please visit us at
wwwcountrysidebooks.co.uk

ISBN 978 1 84674 200 2

*This book is dedicated to the work of beekeepers everywhere
who are striving for the future survival of this incredible species;
to Doug and Lester at Beverley Beekeepers; and to Mum
who always helps me to keep the girls going.*

Designed by Peter Davies, Nautilus Design
Produced through MRM Associates Ltd., Reading

Printed by Information Press, Oxford

CONTENTS

	Introduction	4
Chapter 1	The Honeybee and the Colony	6
	Honeybees, other bees and honey collectors – inside the honeybee colony – the queen – the worker – the drone	
Chapter 2	Cooperative Living and Honey Production	13
	The waggle dance – pheromones – how is honey made? – what should not be present in honey – other hive products	
Chapter 3	The Lifecycle of a Colony	22
	The bee colony in a managed hive – swarming and queen replacement	
Chapter 4	The Annual Work of the Beekeeper	28
	A twelve-month plan – swarm control measures – a note on stings	
Chapter 5	Setting up as a Beekeeper	34
	The modern hive – urban beekeeping – where to put your hive – equipment and clothing – preparing honey for sale	
Chapter 6	Bee Diseases and Related Problems	46
	Bee disease information resources – food related problems – poisoning – parasites of the honeybee – diseases – Colony Collapse Disorder	
Chapter 7	Gardening for Bees	54
	Growing for bees – plant list by month	
	Glossary	60
	Useful Addresses	63
	Index	64

Introduction

Have you ever thought about becoming a beekeeper? More and more people are coming to appreciate the role in our lives of this wonderful creature, and wondering if perhaps they too – even in an urban environment – could welcome bees into their garden or allotment.

This book is designed to be a first step in explaining the honeybee and the work of the beekeeper, both in rural and urban situations. Anyone with time and dedication can become a good beekeeper – a role that will reward you not only with honey, but also with a hugely satisfying pastime.

Bees have coexisted with humans and other animals for millennia, for bees are far older as species than we are. The bee that we have the most contact with in our daily lives is the one that produced the first sugar fix that humans were hooked on – the honeybee, *Apis mellifera*.

And yet, it is this most familiar bee that is one of the most threatened. Few honeybees live as true wild bees today – in the UK hardly any survive outside the relatively controlled environment provided for them by beekeepers.

Changes in agricultural practice, disease and climate change, amongst other factors, have led to a decline in the numbers of both honeybees and their relatives – the familiar bumblebees that we all know and recognise.

The honeybee is unique, though, in that it is the one bee, and one of the few insects, that has been widely domesticated. Several strains of honeybee exist around the world; these are often referred to as races. Each has their own characteristics: some are calm and lazy, others hardworking, and some aggressive, but it is the European honeybee, *Apis mellifera*, that has become man's best friend.

Given their distribution, it is acceptable to suggest that the domestication of honeybees took place gradually and in several places. We know that the Egyptians collected honey to travel with them in the afterlife – and we know that they had domesticated bees because we have images in their tombs of honey gathering.

The European honeybee, above all others, is well suited to domestication because it is hard working and is generally not aggressive. Over the centuries during which man has learnt to harness the many properties of the honeybee, our relationship with the species has become increasingly sophisticated. Where honeybees exist in the wild in large numbers, honey is sometimes still gathered from the cave, rock or tree in which a colony has settled. In some cultures, bees are still maintained in the walls of homes, with easy access to the honey through a special door for the cook.

The development of the modern hive has allowed beekeepers to further increase the sophistication of

Introduction

Gerard and the girls.

and pests with them. It is in the past half century that this movement of disease has come to threaten all honeybees with potential extinction. Careless disease control has meant that today our beekeepers, scientists and policy-makers in government need to work closely to protect not just the survival of the honeybee for its own sake but that of mankind itself.

We must also safeguard the massive ecological and economic contribution that honeybees make to the world of agriculture because of their role as pollinators of the many flowering plants that co-evolved with bees some 100 million years ago.

their art. Prior to this, entry into a traditional skep or woven straw hive in the search for honey would have probably meant the destruction of the colony. A modern bee hive allows the keeper to enter directly into the hive and remove honey or manipulate the bees at leisure. This has proved essential for the advancement of our understanding of the biology of the bee, but it has also led to the many problems facing our honeybees today.

As the art of beekeeping has evolved into the science of beekeeping over the course of the past 200 years, agricultural practices common to other forms of animal husbandry have been transferred to bees, such as breeding for improved characteristics (for this, read 'honey gathering').

The movement of honeybees around the world has occurred for many centuries – from about 1600, honeybees were transported to North America. But, more recently, the movement of bees on a massive scale has led to the transmission of disease

If you cannot become a beekeeper, for whatever reason, you can still help all of our bee species by planting bee-friendly flowers and shrubs in your garden, allotment or window box, and there is a list of plants in Chapter 7 that will help you achieve this.

By understanding the honeybee, and by becoming a beekeeper, it is possible to help both the bee and our wider environment. In the UK, the British Beekeepers' Association supports hundreds of local groups which offer a range of advice for the novice. It is within this framework of training, mentoring and disease control that anyone considering becoming a beekeeper must first approach the honeybee. A coherent approach to our relationship with the bee is the only way that we will ensure its future and our own.

Gerard Baker

Chapter 1

The Honeybee and the Colony

The European honeybee is one of only a handful of bee species that man has domesticated around the world and forms the bulk of those bees that are managed for the production of honey. It has been transported by man from Europe to places where there are no native honeybees – to the Americas, Australia, New Zealand and to the Pacific. Although there are other races of honeybee, the European is the bee most suited to the production of honey within hives. In the UK, this bee now almost exclusively exists in domesticated colonies managed by amateur beekeepers and larger scale bee-farmers because of the effects of disease in wild populations.

The honeybee is the best-known social insect – the name we use to describe those insects that live in colonies. No single honeybee will exist outside the parent colony and the lives of all honeybees are interdependent.

We often refer to a colony or 'hive' of bees because each managed hive of bees will contain just one queen and her progeny, both male and female. Up to 80,000 honeybees may live in one colony, and it is useful in some ways to think of it as a single organism when it comes to later discussions of the colony's reproduction.

Key to the evolution of the honeybee was the co-evolution of flowering plants, which produce nectar, and of those enzymes within the bee that are necessary for the conversion of the sugars in nectar to honey, which can be stored as a winter food. It is this feature of the honeybee which makes it so attractive as a partner to man. The production of an excess of honey, which can therefore be harvested, is the aim of the beekeeper.

Other bees and honey collectors
Three other species of honeybee are recognised – *Apis cerana*, *A. florae* and *A. dorsata* – all occurring in parts of tropical and sub-tropical Asia. Additionally, some types of stingless bee, the *Meliponinae*, contain large species that produce harvestable quantities of honey. Certain wasp and ant species also gather quantities of honey that are harvested by native peoples, mostly in the tropics.

Bumblebees
On a warm spring day, many different types of bee will be flying, looking for food and new nest sites. Those most commonly seen will be the bumblebees and their close relatives – the large and furry bees that we all know and love. About 25 species of bumblebee live in the UK, a number that varies with the migration of species from warmer climates to our shores and also with the loss of bees because of

habitat destruction and the effects of climate change.

Honeybee ecology is very different from that of the bumblebees, none of which survive as entire colonies through the winter. Like the wasp, bumblebee colonies die in the autumn with only new queens hibernating through the winter and emerging in early spring to begin a new nest. Bumblebees are effective pollinators for plants which are unsuitable for honeybees because of the long length of their flower tube, as honeybees have relatively short tongues.

Inside the honeybee colony

Before we go on to look at the structure of the physical hive, let's look at the bees that are found in an individual colony. Throughout the year, the composition of the colony will change. During the spring and summer periods of growth and development there are both male (drone) bees and female (worker) bees in addition to the queen. Young bees in varying stages of development are present in much larger numbers between March and October than in the colder months and the presence of food (nectar and pollen) within the hive varies with its availability during the year. A colony without food will die if none is available.

THE QUEEN

Each honeybee colony is the production of one adult female queen, the largest bee in the hive

A queen cell is noticeably larger than the surrounding worker cells and hangs vertically from the honeycomb.

and the product of a fertilised egg. The development of a new queen within a colony is determined by the colony's instinct to swarm and divide into two or more colonies. In normal circumstances, swarming is how honeybee colonies divide and multiply.

From egg to queen
Large queen cells are produced by workers. These stimulate other nurse bees to feed the larval queen a different mix of secretions to ordinary female workers. A highly nutritious food, royal jelly, and foods with a high sugar content stimulate the development of the female egg into a queen in the first three days of the life of the larvae.

When ready to pupate, the queen is sealed into her cell, uniquely with additional food, and emerges approximately 16 days after the egg was initially laid. The emerged queen will not initially be able to fly, and for the first few days is light-phobic.

She will, however, interact with other bees within the hive, moving quickly from comb to comb to check for other queens. Any that she finds will be stung to death, and she will often harass workers, who in turn will harass the queen. Worker bees will repeatedly chase the queen around the hive and follow this by feeding her – a process which leads to increased fitness and vigour. Fitness is necessary for the queen to sustain her maiden flight and safe return to the colony once mated.

The mating flight
A week or so after emerging, the queen will embark on her maiden flight, during which she will attempt to mate with male bees known as drones. Flying out of the hive, she will spiral up to a height generally above that at which normal workers fly and most likely on days that are warm and calm. Mating may occur with several drones, whose sexual organs are literally exploded into the queen.

She is unlikely to leave the hive again until her colony is large enough, or needs, to swarm. After maybe one or a small number of mating flights, the queen will not mate again and is able to store the live sperm for several years in a special organ, the spermatheca.

Having mated, the queen will lay productively for two or three years. She is capable of laying both fertilised and unfertilised eggs. The fertilised ones combine her genetic material and that of her mate, and these eggs develop into female worker bees; unfertilised eggs are laid in small numbers and these develop into male drones, which have little purpose in the hive other than the role of being a mate to a newly-emerged queen.

How are queens replaced?
A queen may be superseded within the colony for a number of reasons. The colony is held within a tight balance between, on the one hand, the chemical control of the workers by the queen's pheromones, and on the other evolutionary pressure that causes the workers to rid themselves of an under-performing queen that has declined in health or ability to lay fertilised eggs in sufficient quantity.

The queen produces 20 or more pheromones that affect the behaviour of the other bees in the hive, including those that control the development of ovaries in the worker bees, preventing any question of competition to her role within the colony.

The Honeybee and the Colony

A failing queen will lay drone brood. Here, worker cells have been enlarged to contain drone brood. This colony swarmed, replacing the old failing queen with a new generation.

Pheromone production from the mandibular glands of the queen appears to be strongest when the young queen is mated and in what we might refer to as her prime laying period. As the quantities of these pheromones decrease, either because of age or loss of vigour, worker bees become more likely to replace her with a new queen. So long as there are female eggs present in a hive, workers are able to rear a new queen.

Workers

More than 90% of the bees within each colony are worker females, the result of a fertilised egg laid by the queen. The food fed to the larval worker is different from that fed to queen larvae and determines its development.

Workers perform almost all the roles within the hive: they nurse and feed the developing brood, build honeycomb, forage, share information

Bees & Beekeeping Explained

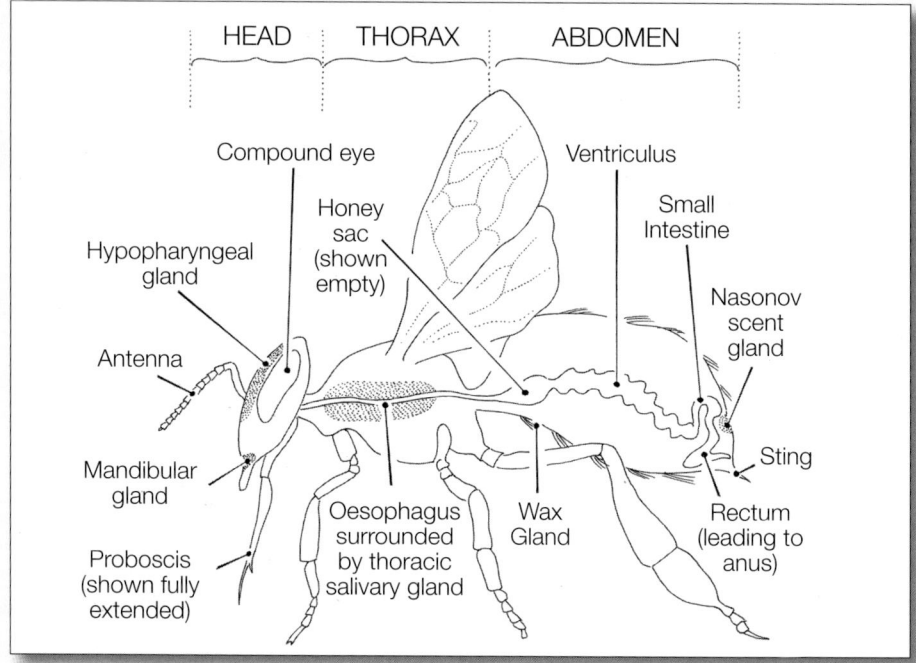

Section through a worker honeybee.

about their environment and defend the hive. Rarely, and usually when a queen is absent, some workers develop the ability to lay unfertilised eggs.

Feeding the growing colony
The queen lays worker eggs in numbers that correlate positively to day length and the availability of food in the environment. Workers are fed secretions of food that differ from those fed to the queen, containing noticeably less sugar.

The larval bee is a creamy white, maggot-like grub, which is fed by other workers for the first five days after the egg hatches. A massive spurt of growth is achieved, and in just five days the grub moults four times until it has increased its weight by two thousand times, at which point it is sealed into its cell. The grub then spins its cocoon and pupates into the adult worker.

The female honeybee emerges some 20 or so days from the egg being laid, and once emerged feeds mostly on pollen and quickly reaches its full adult size.

The worker bee's many roles
When food is available, worker bees gather in nectar and pollen from flowering plants, and water as required. They also occasionally utilise other food sources such as

The Honeybee and the Colony

aphid honeydew for use by the young developing brood and to store for periods of food scarcity.

Worker bees are able to survive on undiluted nectar and do not need to ingest water. However, water is required by nurse bees to dilute the honey they use to feed the developing brood.

The life of a worker bee in the summer is a matter of only a few weeks, and yet within that time she will undertake a large variety of tasks.

During the first few days of adult life, the pollen ingested by the worker provides the protein necessary for the development of her hypopharyngeal and other glands that secrete brood food, royal jelly, wax and venom,

On warm days in spring, bees will fly out to defecate.

and the enzymes necessary for the conversion of nectar into honey. Her early days are spent on the brood comb, cleaning empty brood cells and maintaining the brood-rearing capacity of the colony.

From colony maintenance, the worker moves on to wax and honey production, thence to colony defence and finally to foraging for food and water for the last few days or weeks of her life.

During the summer, the life of a worker may be as short as five or six weeks. The workers born in the late autumn are able to survive the winter for several weeks longer than this, due to various physiological changes that occur within their bodies that enable them to store more food internally.

Worker bees are crucial to almost every aspect of the life of a colony, including the very temperature within the hive. By fanning their wings at the hive entrance, or by contracting their decoupled flight muscles, the bees can lower or raise the temperature within the hive to a considerable difference to the temperature outside. The internal temperature within the hive hovers around 35°C during brood rearing. It is less in the winter.

Drones

A summer-time colony will contain just a few hundred male individuals (drones), each the result of an unfertilised egg. The queen is able to determine which eggs are fertilised as they are laid and, as such, can limit the number of males in the hive. Should she run out of stored sperm in later life, she will become a drone-layer and will most likely be superseded by the workers.

The drone bees' central function is to mate with a new queen, upon which they die, having exploded their genitals into the virgin queen.

The development of the drone takes place in a similar way to that of the workers, but the larval drone is fed for seven days before pupating, and takes longer to develop in its cell than both types of female honeybee. Drones are therefore more susceptible to attack by certain parasites than other bees, a subject we will come back to in Chapter 6.

So long as there are other colonies of bees nearby, drones may mate with virgin queens from other hives, maintaining hybrid vigour in the wider population of bees.

It is known that drones congregate in certain zones about 15 metres to 25 metres in extent, which are often above the same piece of ground from year to year. It is not known why these zones attract bees.

The drones attract each other and virgin queens by secreting a pheromone from their mandibular glands, and are similarly attracted to a pheromone secreted by the young queen. This results in a virgin queen being mated by a number of males. If colonies are isolated, it is possible for bees to become inbred.

Because of the tendency of drones and queens to congregate together in such zones, the virgin queens are particularly susceptible to predation at this time.

Drones do very little within the hive. They do not forage for food or water, do not tend other bees, and are fed by female workers for their, average, three or four weeks of life. Late in the season, they are starved and thrown from the colony as the female workers ready the colony for the winter.

Chapter 2

Cooperative Living and Honey Production

A blackthorn hedge in flower – good foraging for the early bee in good weather.

The ability of honeybees to communicate and cooperate has been the key to their ecological profile. Although there are approximately 19,000 different species of bee around the world, none has the ability to out-compete the honeybee for resources, a dominance only threatened by predator species, climate change and disease.

It is thought that all bees evolved from wasps about 100 million years ago, roughly at the same time as flowering plants, the *Angiosperms*, began to emerge. If we consider that single plants may have been dispersed throughout a primitive forest environment, it is easy to see how individuals who could communicate the position of such a food source would have an advantage over rivals who could not.

Chemical signals and physical communication in the form of dancing are used by scout honeybees to communicate not only that there is food available locally but, crucially, where the food source is in relation to the colony and the sun.

Communication

THE WAGGLE DANCE

A worker honeybee returning to the hive is able to communicate with great accuracy where she has located food, and recruit others to find the food source. If the food source is near to the colony, the taste and smell of the nectar and pollen the successful forager has brought back to the hive will provide some information to fellow workers. If the food source is within 30 metres of the hive, a simple excited round dance will encourage other foragers to leave the hive and search for food.

Bee communication reaches the peak of its sophistication, however, in the form of the waggle dance, largely decoded by the work of the pioneering biologist Von Frisch in the 1960s.

The waggle dance is a ritualised version of the bee's successful foraging journey. In it she will communicate the nature, direction and distance of the food source from the colony's home.

Honeybees use the position of the sun and its polarised light to orientate themselves, so the bee dances in a way to suggest the direction of the food in relation to the sun's position. On a vertical section of comb, the bee dances in relation to gravity – if the dance is vertically aligned on the comb, then she indicates that the food source is directed towards the sun.

If, on the other hand, the food source is located at an angle of, say 40° to the left of the sun, the dancer will orient her body 40° to the vertical on the surface of the comb (see diagram below), and will repeatedly circle around, wagging her abdomen, an action which recruits other worker bees who are able to memorise the

The waggle dance performed by a successful forager gives information to fellow workers, enabling them to locate a food source.

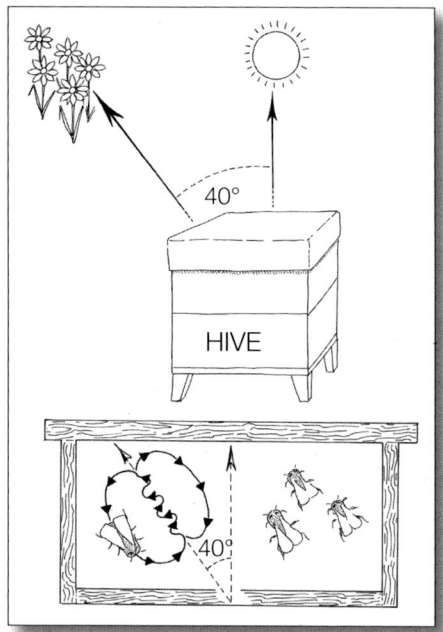

Cooperative Living and Honey Production

orientation and find the food source.

In addition to the direction of the food source in relation to the sun, other information is transmitted through the dance. The distance of the food from the colony is indicated by the frequency of the dances – the nearer that food is to the colony, the quicker the bee dances.

Simultaneously, the bee emits sounds that also indicate the distance of the food to the colony. Sound and vibration are important in this communication, as the interior of the colony, or hive in the case of domesticated bees, is dark.

The bees following the dancer are able to interpret the dance using vibration and sound and in return use auditory signals to communicate to the dancer. Using a 'stop' sound signal, they encourage the dancer to pause and pass to them a sample of the nectar or pollen collected, which tells them the flavour or scent of the pollen and flowers they need to locate on their own foraging journey.

Interestingly, pollen has a smell that is different to that of the flowers that produce it, adding another layer of complexity to this method of communication.

Pheromones

Pheromones are produced by various glands in queens, workers and drones. These are chemicals designed to have an effect on other bees by being transmitted through the environment. In addition to those glands in the head such as the hypopharyngeal and mandibular glands, the worker bee possesses a gland on the dorsal side of its abdomen, the Nasonov gland which it uses to attract other workers. Typically, at the hive entrance, bees will be seen with their rear ends raised, exposing the Nasonov gland. Other bees so attracted may then engage in whatever behaviour is stimulated – be it feeding, defence or attack.

How is honey made?

Since biblical times, man has exploited the honeybee's tendency to collect nectar and process it into honey. It is not certain whether man first domesticated bees to provide honey or to pollinate their food crops – happily the two go hand in hand. Foraging bees collect both pollen, as their main source of protein, vitamins and minerals for young bees, and nectar as a source of carbohydrate food energy. Bee products in addition to honey are discussed later, but for now let's look at how nectar is converted to honey.

Flowering plants produce nectar to attract pollinating insects, and many thousands of insect and other species fulfil this role, with bees, wasps, flies, butterflies and moths being the most common groups. Honeybees differ from almost all other temperate insects though, in that they exist as large colonies throughout the winter, and therefore require a store of food to keep the colony alive during periods of food scarcity.

Although honey and pollen can be stored within the hive, water is only able to be stored in the bodies of the bees themselves. Water availability is essential for the temperature control of a colony as well as for feeding the brood.

When honeybees collect nectar from flowers, the nectar is not in a suitable state to store because its sugar content is too low to prevent moulds and yeasts spoiling it. It must therefore be concentrated and converted into honey.

Bees & Beekeeping Explained

Different species of flowering plant secrete different amounts of nectar from organs called nectaries in their flowers. Nectar is a modified form of plant sap containing a mixture of three sugars – sucrose, fructose and glucose in varying proportions depending on the species of plant.

Honeybees will also at times collect plant sap from extra-floral nectaries, and honeydew that is exuded from aphids and which may therefore originate from non-nectar producers such as grasses. Although honeydew so collected can be a useful food source for bees, it can only be marketed as such, and not sold as honey.

Weather conditions obviously determine how much nectar can be gathered by honeybees, but not just because low temperatures or rain will limit foraging behaviour. Plants produce nectar in response to climatic factors, generally the more nectar the more sunshine they receive. Humidity and lack of available water in the soil also affect nectar production so that some plant species will produce more honey one year than another.

As might be expected, beekeepers will try to locate their hives near to plants that generally offer a high potential honey crop for their bees, leading to a migration of bees from crop to crop at different times of year. Urban beekeepers, on the other hand, may find that the gardens surrounding their apiary offer such a mix of flowers throughout the year that their hives never need to be relocated.

In many countries, borage is grown for its seed oil. The light, flavourful honey is popular amongst beekeepers.

Cooperative Living and Honey Production

Bees & Beekeeping Explained

The queen excluder sits between the brood chamber and the honey 'super' and is designed to prevent the queen laying eggs amongst the honey, which would make extraction difficult.

How the Bee Makes Honey

The honeybee uses its tongue to collect nectar that is then stored in the first part of the gut, the honey sac, where the bee filters out any pollen grains, bacteria and fungal spores.

Key to the production of honey is the enzyme invertase, which originates in the hypopharyngeal gland in the bee's head and which begins to work almost as soon as the nectar is ingested. Invertase breaks down any sucrose in the nectar into fructose and glucose. The temperature in the hive allows the solution of sugars to be much more concentrated than at higher or lower temperatures, so the solution of fructose and glucose can be 80% or more, resulting in a honey that can be stored.

When the foraging bee returns to the hive, she must concentrate the nectar she has gathered, a process that results in its conversion to honey. Positioning herself on the comb, she exudes the nectar on her tongue, and repeatedly ingests it.

By exposing the nectar to the warm hive interior, water is evaporated,

Cooperative Living and Honey Production

thereby concentrating the solution of sugars. The worker repeats this, adding more salivary enzymes to it as it dries, before placing it into a storage cell in the honeycomb.

Here, it may be further concentrated by evaporation (often by bees beating their wings over its surface to draw off moisture) until the bees deem it to be of sufficient concentration to store without spoiling.

The proportion of glucose and fructose within the honey determines whether the honey will remain liquid (high fructose) or granulate (high glucose). Generally, honeys contain slightly more fructose than glucose, but some contain notably more glucose, including oilseed rape honey which is common in the UK. Because fructose is a particularly sweet sugar, sugars with high fructose content will naturally taste sweeter.

Other Constituents of Honey

Apart from the sugars referred to above, honey contains other substances, including about 20% water. Additionally, a small amount of pollen, some proteins and amino acids and minerals may be present.

If the water content is high enough, any yeasts present in the honey will cause the honey to ferment. Some invertase may also be present in the finished honey. In unripe honey another enzyme, glucose oxidase, may be present which produces gluconic acid, the main acid element of honey, and the antimicrobial substance, hydrogen peroxide.

The completed honey, therefore, is a viscous, sweet substance with a high osmotic pressure, a low pH and antimicrobial characteristics, and all of these features make it more likely to store without deteriorating.

What Should Not Be Present in Honey

Because honey generally has a monetary value higher than that of sugar, it has often been adulterated with sugars in an attempt to defraud the buyer. Starches and sucrose from maize and sugar cane can be converted into glucose and fructose, but methods exist to detect such adulterations.

Occasionally, honey claiming to be home-produced has been found to be mixed with cheap honey produced overseas, notably from South America and China. The study of pollen grains present in honey can detect this and prosecutions are often made by Trading Standards authorities.

Some plants produce toxic constituents in their nectar or honeydew which can be transferred to honey, but examples are rare, while some such as euphorbias produce honey which can cause a burning sensation, but this is rarely a problem in the United Kingdom.

Honey is not recommended for young infants because of the occasional presence of botulism spores. This appeared to be a particular problem in California in the 1970s as their soils contain such a high amount of this bacterium.

Occasionally, honey may contain substances used to treat bee diseases as well as pesticides from nearby crops. It is the duty of the beekeeper to ensure that the honey they are producing is not subject to contamination of any kind.

Other hive products

Wax

Beekeepers use sheets of beeswax imprinted with hexagons that are the perfect size for worker brood cells. When given to a colony of bees in a hive, the bees are stimulated to form hexagonal columns that are roughly perpendicular to the vertical sheet – thereby forming storage cells in which young bees can develop and in which food can be stored.

Young bees produce wax from glands on the ventral side of their abdomen. In time, this wax becomes old and stained, and is often replaced by beekeepers. The result of this continual renewal is a source of wax which may be used to manufacture candles and polish.

Cooperative Living and Honey Production

Propolis

Bees collect propolis, a resin-like substance, from certain shrub and tree species. The resulting substance is plastic and flexible when warm, hard and brittle when cold. It can be used to fill gaps in the hive, and is often used by the bees to coat foreign substances in the hive such as dirt particles, even dead mice. It has anti-microbial properties and is sought for adding to creams and tinctures.

Pollen

A highly nutritious food that can be collected from bee hives through the use of a pollen trap. A large colony will collect several pounds of pollen throughout the season, although of course they collect it principally to feed the growing colony. Pollen is used in some areas as a health food.

Royal jelly

Royal jelly is collected from queen cells and is produced in very small quantities within the hive. Colonies can be manipulated to produce more than normal, but its production and collection is a specialist art.

The bees working on this super frame have propolised the top surface of the bar – orange propolis stands out against the pale wooden background.

Chapter 3

The Lifecycle of a Colony

This section of the book considers the lifecycle of the bee colony in the context of a managed hive – the name that we give to the physical structure in which beekeepers keep a colony of bees, usually headed by one queen.

Throughout the year, the hive will contain adult bees, eggs, and young developing bees known as brood. During the winter months, or months of food shortage, there will be few eggs and brood. In the spring, as day length increases and food in the form of pollen and nectar becomes more available, the queen is stimulated into laying more eggs. At her peak a queen is capable of laying over 1,000 eggs per day.

Given unrestricted food and water availability, a colony will soon fill the hive. In practice, food availability does vary, as does the genetic make-up of each colony, so that it will take up most of the available brood-rearing space in a hive at slightly different times. In the UK, this happens between May and July. If the colony has no further space to expand into, the workers will prepare to swarm even earlier than this.

A beekeeper will plan the year to have their hives at maximum strength, i.e. full of bees, when the crops they are alongside are producing their maximum honey flow.

Traditionally, a beekeeper will manipulate the biology of the bees in their colony to prevent swarming, which would otherwise mean that the bulk of the flying bees – the honey gatherers – would be lost at the period of maximum honey flow.

Organic and biodynamic methods of beekeeping advocate allowing the bees to swarm, and attempting to then catch the swarm to increase the number of colonies in their apiary. This may result in little honey being available to gather, but an increasing number of people are keeping bees 'organically' purely to maintain the species.

Chapter 4 deals with the management of the hive month by month, but first let's take a quick look at what happens when the colony readies itself for swarming.

Swarming

When a colony swarms it divides into usually two, maybe more, parts. Swarming will usually occur when the colony is in its biggest growth period – when it fills the available space in a hive and therefore becomes crowded.

However, there are several reasons why a colony may swarm – it could be because the queen is old and therefore

A small swarm begins to form near to the original hive in June.

The Lifecycle of a Colony

Bees & Beekeeping Explained

The Lifecycle of a Colony

The swarm and attached foliage are lowered onto a brood box full of drawn comb – the bees will move down towards the dark and take up occupancy in their own home.

not laying well or not producing enough of those pheromones that normally prevent the workers producing new queens.

Replacing an established queen

As the workers ready themselves and the queen to leave the hive, several new queens are being developed, one of which will take charge of the bees that remain when the swarm has departed. The development of a queen is discussed in Chapter 1.

The bees that remain are only those who are too young to fly – these are needed to feed the remaining brood and look after the developing queens and go on to become the basis of the new colony.

The process of making ready to leave the hive takes place while the larval queens are developing in their queen cells.

When the workers begin to construct new queen cells and develop new queens, other changes occur within the colony to make ready for swarming. The queen is fed less, will lay fewer eggs and lose weight so that she is ready for flying out with the swarm. At the same time, workers will keep their honey sacs full so that they take food with them when the swarm finally issues from the hive.

Finally, a number of worker bees also run throughout the hive, shaking their wings and actively shaking the queen, finally almost forcing her from the hive.

Bees & Beekeeping Explained

The swarm moves down into the hive.

The swarm departs

The queen larvae are sealed into their cells to metamorphose into adults around eight days after the eggs were laid. At or around this point, the swarm will issue from the hive if the weather is suitable. If it is not, they may wait, even delaying the emergence of the new queens until the swarm can leave the hive and establish itself in a suitable new place.

If there are several queen cells left in the hive to emerge, one or more of the new queens may emerge with another swarm, further reducing the number of bees within the hive to perhaps only one quarter or one third of the original.

Key to the success of a colony that swarms is that both the swarm and the small colony left behind have enough time in the remaining year to develop into a colony strong enough to survive the following winter.

The virgin queen that emerges to take charge of the remaining colony will first check to see if there are any remaining queens, and will usually sting them through the walls of their cells, killing them.

Her pheromones will soon begin to control the workers in the hive, and by making a piping sound that is created by pressing her body against the comb and vibrating her closed wings, she indicates her presence to her new colony.

Although swarming usually occurs

The Lifecycle of a Colony

in the early summer, a second period of swarming can occur in the late summer if food availability has allowed the colonies to build up large numbers of bees.

Swarms should not be feared

A swarm of bees issuing from a hive is one of the most remarkable sights in nature. The noise and sheer number of bees – tens of thousands – can be daunting, and yet swarms present little threat to the public and should be left alone to settle and calm.

Before departing from the hive, workers have filled up with honey. In addition, they are intent on only one thing – following the queen and establishing a new colony. They are therefore less inclined to sting or attack unless provoked. It is rare for European honeybees to attack en-masse, unless their colony is directly attacked.

Initially, a swarm will settle near to the original colony site, perhaps in a tree or shrub where they may remain for several hours or days. From here, scout bees will search out a new nest site. Beekeepers frequently place empty, bait hives in their apiaries to attract any swarms that may derive from their own hives or be passing from a neighbouring keeper.

Generally, it is assumed that if a keeper finds a swarm of bees, he is entitled to keep them as long as the landowner agrees. In practice, many keepers would return a swarm to a fellow keeper if it were obvious that the swarm had issued from his hives, although of course, it is nearly impossible to prove the provenance of a swarm at swarming time.

Making ready for winter

After swarming time, the colony continues to build up numbers and make ready for the winter. A colony of bees requires food stores to survive over the colder months. Beekeepers will remove honey from a hive throughout the summer season that is in excess of that required by the bees themselves.

Traditionally, beekeepers will remove all surplus honey from the hive, and replace what they take with sugar syrup. Increasingly, so-called organic methods are being followed which advocate only removing honey in the spring and leaving the colony to develop its own winter food store for the rest of the summer.

To survive the winter, a mature hive of bees will require approximately 20 kgs of honey. Should the colony have not been able to collect enough stores to ensure strong development or survival, the keeper will need to supplement their food. It is essential that bees going into winter have sufficient food stores. Winter feeding is discussed in Chapter 4.

The bees within the hive cluster in the winter and reduce their metabolic rate to reduce their food consumption. They are stimulated to do this by external air temperature: normally the bees would be dispersed throughout the hive, but below about 14°C they begin to form a distinct cluster that contracts as the temperature dips.

The bees on the outside will naturally be colder and so they move inwards and change places regularly. The larger the cluster, the easier it is for the mass of bees to maintain a temperature that will enable them to survive.

CHAPTER 4

The Annual Work of the Beekeeper

T he beekeeper's work runs year round because even though honeybees may cluster and shut down to some extent during the winter, the beekeeper must check that the colony has constant access to food. From the shortest day, egg laying will recommence and therefore food stores within the hive will be consumed at a faster rate.

This chapter discusses the annual work of the beekeeper, giving a month by month guide to which tasks should be completed with the bees. I will go on to discuss what equipment and training you will need to begin beekeeping – and where to site your hive – in Chapter 5.

The acquisition of bees will normally occur in the early summer – the period in which bees are most usually available.

One of the few surviving hives from the grand age of Victorian country house beekeeping. This Carr-Stewarton hive of 1863 has woven wicker walls, a copper roof with finial, and a viewing aperture for checking temperature and humidity. (From the collection of the late Murray Armor, currently housed at Hodsock Priory, Nottinghamshire).

JUNE

STARTING OUT WITH A NEW COLONY OF BEES

Most beekeepers start their keeping life by obtaining a small colony, called a nucleus of bees, from a fellow bee-club member or by similarly obtaining a swarm. Either way, the new bees will need help in getting established.

Before your bees arrive, you should be prepared for them. Ensure that

The Annual Work of the Beekeeper

you have the correct equipment for safe handling, discussed in Chapter 5. Whoever you are collecting the bees from will be able to help you transfer the bees into your hive.

A five-frame nucleus will take up half the new hive, and the frames of bees should be carefully placed in the centre of the new hive, with empty frames containing just foundation beeswax either side of them for the bees to expand into. On no account should you place empty, undrawn, frames between frames containing developing bees, i.e. brood, as these are easily chilled.

Once you have transferred the frames into the hive, place a queen excluder on top of the brood chamber and a super on top of this with the frames of wax aligned in the same direction as those in the brood chamber beneath. A crown board can then be placed on top of this.

To feed the bees, place an empty super on top of the brood chamber and place your feeder over one of the holes in the crown board. Do not cover both holes. Place the hive roof on top of the whole stack and place a brick on top of this for extra stability.

Feeding a nucleus
A small colony will need to be fed to enable it to establish quickly. Do this with dilute syrup, using only white sugar, in the proportions of 4 kilos of sugar in one gallon of water in a contact or bucket feeder. The food will provide the necessary energy for the developing colony to draw out the wax in the hive into honeycomb for brood rearing and food storage.

Various types of pollen substitute are also available and can be essential if the local pollen availability is low at any time.

Apart from checking for food, colonies should be allowed to settle in a new hive for two weeks before you re-enter the brood chamber.

For an established colony

Your colony will be building rapidly if food is available, and will be at risk of swarming. Check the brood chamber every eighth day for signs of queen cell building and knock down any that appear. There are various methods of swarm control available to you, and these are discussed at the end of this chapter.

If the colony is struggling, and little or no female brood is being laid, check that the queen is still present and make plans to replace her as soon as possible.

Disease management

Once the colony is settled into a hive, repeated checks for disease should be made. For discussion on the methods of identifying diseases, refer to Chapter 6.

Disease monitoring should take place each time the brood chamber of a hive is opened to check for swarming activity, and preferably once every eight days during the warmer months. Treating for disease should be carried out whenever necessary, but it is essential to follow accurately all instructions given by the manufacturer of any chemical treatment.

Colony development

As the colony develops in the summer, the pattern of brood laying will move from the centre of the hive outwards until all of the brood frames are occupied with the process of brood rearing, which will reach its peak in July.

JULY

For new colonies

Check that the brood chamber has begun to fill out and continue to feed as appropriate. As long as the queen is continuing to lay, brood will be reared and the colony will expand quickly.

For established colonies

If your colony has built up strongly, you will be in the happy position of being able to extract honey. See the guide to equipment in Chapter 5 for advice on what you need to do this.

Add additional supers to the hive as necessary to give the bees plenty of space in which to store honey. Consider adding a super between the brood chamber and the queen excluder, giving more space for the colony to expand their brood rearing capacity.

AUGUST

For new colonies

Check that your colony has fully drawn all of the comb in the super – if it has, stop feeding and allow the bees to continue building until you need to feed in the autumn.

For established colonies

Make sure that your brood chamber is still filling nicely. If the bees have not made any attempt to create new brood in the additional super you added last month, remove it and allow them to continue placing honey in the frames above the queen excluder that will form the basis of their winter food store.

Disease management

Make sure that the bees are still free of disease, and if not refer to the guide to disease management in Chapter 6.

SEPTEMBER

For new colonies

Ensure that the colony is going to have enough food for the winter. Look into

The Annual Work of the Beekeeper

the super during the first week of the month and see if there are any capped stores left in there.

If the super is less than two thirds full of honey, feed a strong syrup – 1 kilo white sugar dissolved in 500mls water – up to the first week in October so long as the weather is mild. Using a bucket feeder, you will need to check a couple of times a week to keep the feeder full. Check, too, for any sign of mould in the feeder and wash well if any appears.

By gently feeling the weight of the hives in winter, a beekeeper determines whether or not the bees have enough food stores.

For established colonies

Remove any surplus honey and feed as above until the colony has 20kgs winter stores available. See the diagram of a hive (page 36) for where to place the feeder. This position helps prevent cooling of the hive when you are accessing the feeder to top up.

DISEASE MANAGEMENT

If your *Varroa* monitoring indicates that there is a large population of mites present, treat appropriately. See *Varroa* management techniques in Chapter 6.

OCTOBER

FOR ALL COLONIES

Remove all syrup feeders as the bees will be unlikely to ripen and cap any more stores. To be certain that there is enough food, add fondant (sugar paste) to the hive in order to make absolutely sure that there are emergency food rations.

To use fondant, place a kilo of fondant in a freezer bag and stab some holes in the undersurface. Place the pierced surface on top of one of the holes in the crown board immediately over the centre of the frames. The bees will find the food if they need it.

DISEASE MANAGEMENT

Remove any pyrethroid *Varroa* treatments at the end of their time; do not leave chemicals in the hive longer than the manufacturers recommend.

NOVEMBER

FOR ALL HIVES

Make sure that the hives are secure against bad weather. Check for the shifting of hive parts after gales and for fallen tree branches too. Make sure that there are no gaps in the stacks of hive parts.

DECEMBER

DISEASE MANAGEMENT

December is the time of year when there are fewest bees in the colony. Brood rearing declines with day length, so now is the time to treat with oxalic acid for *Varroa*. The reason for this is that at midwinter there should be very little brood in the hive. As oxalic acid does not penetrate beeswax, any mites contained in the cells with developing brood would not be killed.

Oxalic acid is dangerous to handle, but it is possible to buy ready-made and measured phials for dribbling into the hive at midwinter, which simplifies the process of administering it.

JANUARY, FEBRUARY AND MARCH

FOR ALL HIVES

The early months of the year represent the time when most failing colonies die out. It is essential to check that the colonies have sufficient access to food and water. Fondant can be used to some extent, but as the weather warms into March, water can become a problem if temperatures are very low. A dilute syrup of 1 kilo to 1 litre of water can be used in a bucket feeder to provide the growing colony with enough energy.

Pollen substitutes may also be used to provide a much needed source of protein at this time.

Swarm control methods

The simplest and quickest way to avoid imminent swarming in a colony is to remove all incipient queen cells

The Annual Work of the Beekeeper

– those that are being shaped by the workers and which soon will be filled with royal jelly and an egg.

This can only ever be a stop-gap measure though, and a colony intent on swarming will do just that. Advocates of organic beekeeping suggest that colonies should be allowed to swarm, which is after all a natural part of the bee's ecology.

Many beekeepers allow their honeybees to develop queen cells and then divide the colony into two portions before the queen cells are sealed: the first contains the old queen and four or five frames of brood with no queen cells present. This is placed on the site of the original hive. Over the next few days, all of the worker bees will fly into this hive, leaving the new queen cells and young bees alongside.

The second half of the hive is taken to one side, perhaps 1 metre away. In this, the new queen cells are reduced in number to just two or three.

Once these new queens have emerged, the keeper may then decide to keep both the original and young queens. Alternatively, both colonies can be kept and re-queened with bought queens of a specific type, or the best queen can be chosen once the new queen has displayed her laying tendencies, and the two hives condensed to make a large colony that will go on, hopefully, to be strong for the remainder of the honey flow.

A note on stings

Beekeepers will inevitably get stung by their bees, but this need not be a regular occurrence. Some strains of bees are noticeably calmer and less likely to sting than others – the Carniolan bee is particularly sedate, and yet hardworking. A keeper should not tolerate aggressive bees, for their own sake and that of their neighbours.

Stinging is the last resort of the bee because if the sting is fully inserted into mammalian skin, its barbs pull the internal organs from the abdomen and the bee dies. It must be remembered, when considering why this should be the case, that bees evolved their sting well before mammals appeared on the planet, and honeybees are more than able to repeatedly sting another attacking insect, such as a wasp.

Stings trapped in the skin should be quickly scraped out with a hive tool or finger nail, and any serious reaction treated appropriately with antihistamine. Large, raised areas that are red and itchy should be seen by a doctor.

Multiple stings are unlikely, and are more probably the result of rough handling than anything else. A queenless colony will be more likely to sting than a queenright one.

It should be noted that a sting pheromone is released when a bee stings, so more bees are recruited to attack. Closing the hive and allowing the bees to calm with a little smoke is often preferable to continuing to work in an agitated colony.

So-called 'Africanised' bees have received much press over recent years. The result of the escape of African queens in Brazil in the 1970s is that the African/European hybrids have been particularly aggressive and have spread throughout South Central America and the Southern USA. They are not present in the UK.

CHAPTER 5

Setting Up as a Beekeeper

A beekeeper demonstrates his art in a bee cage at an apiary open day.

Setting Up as a Beekeeper

What steps do you need to take when setting yourself up as a beekeeper? First of all, anyone considering becoming a beekeeper should attend a training course or work with another keeper through their first year. This training is essential if you are to begin to understand the importance of disease monitoring and swarm control. It is also useful, of course, in helping you get a feel for whether you would make a good beekeeper or not, and whether you will enjoy working with bees. This chapter discusses the training available and what equipment is necessary when setting up your hives.

Demonstration hive at a Beverley beekeeper's open day.

Bees & Beekeeping Explained

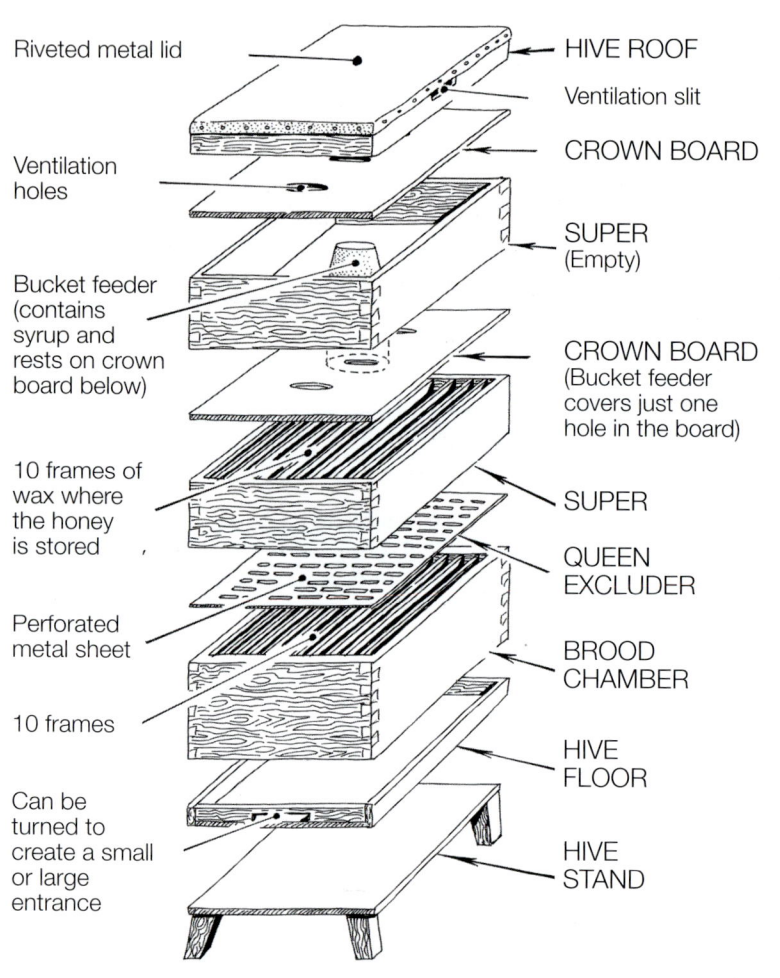

A beehive 'exploded' to show the usual arrangement of parts.

Setting Up as a Beekeeper

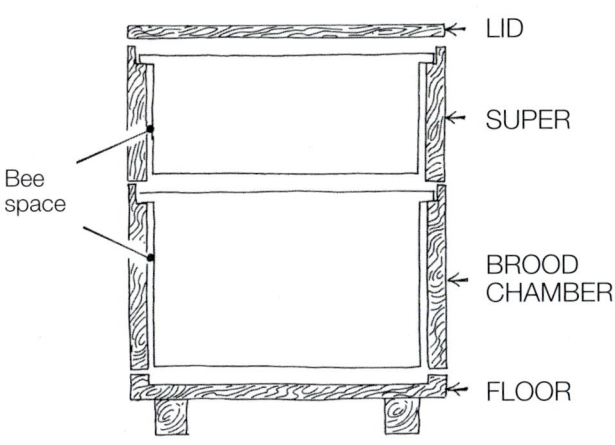

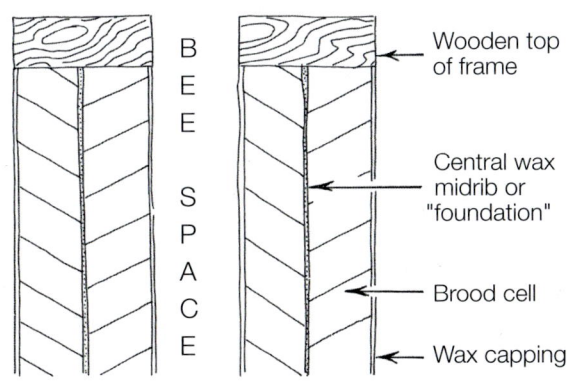

A hive is constructed to exact measurements, preventing any build-up of honeycomb within the 'bee space', i.e. the space between the frame and the hive wall.

Training

Once you have made the decision to become a beekeeper, you should join your local beekeeping association and become a member of the British Beekeepers Association (BBKA). Doing this will put you in touch with other beekeepers in your locality, and also introduce you to the organisation which represents beekeepers at a national level.

The BBKA provides regular updates on all aspects of beekeeping and disease management in particular. Usefully, it also provides bee disease insurance.

A training course may take place over a weekend, or perhaps on one evening a week for several weeks during the beekeeping season. You should aim to have all of your initial training complete before you purchase your first colony of bees. The advantage to this is that you will have had the opportunity to handle bees before you become responsible for your own, and you will have made contacts within your local beekeeping community who will be on hand to give advice when you experience difficulties.

The modern hive

Clay pots, woven baskets and hollowed out logs have all been traditionally used by peoples across the world to provide a honey harvest from various types of bee. Although attempts had been made to create a hive with movable frames, it was not until 1851 that this was finally achieved by the Rev L.L. Langstroth in the USA. He realised the benefit in having frames with a top bar whose lugs extended onto the sides of the hive, so suspending a frame which could be lifted from the hive and reinserted.

Similar hives had been made earlier but the bees' tendency to fill in gaps with brace-comb and honeycomb meant that the frames would become fixed by wax and propolis.

Significantly, though, Langstroth realised that the bees would not attempt to build comb between the frames and the hive walls if the spacing was very specific. This is the so-called 'bee space' and is set by the bees themselves when they construct comb in the wild (see diagram on page 37). The spacing allows bees to work between the combs and minimises the loss of space. Langstroth's hives are still in use around the world.

Various versions of the movable frame hive exist today, and indeed hundreds of varieties of hive are in use worldwide. Whatever the type, the hive is constructed to maximise the ease of management, cleaning and operation.

Wood is the material most commonly used to house bees and has many advantages – it is insulating, natural and can be torched to sterilise it. Other materials are used, such as cardboard, polystyrene and plastic, but are less easy to sterilise.

In the UK, the Modified British National is the most commonly used hive type. The sheer numbers in use make transferring hive parts between keepers and the sale of secondhand hives a simple matter.

Urban beekeeping

Although the information on siting a hive in this section applies equally to a rural beekeeper, keeping bees in an urban environment is increasingly popular. A town or city will have

Setting Up as a Beekeeper

many gardens and parks and is likely to have an increased species diversity over many parts of the countryside, especially arable land growing maybe only three or four species of crop annually.

Urban bees also benefit from the 'heat island' effect – that is, that the ambient air temperature in cities and large towns is often several degrees above that of the surrounding countryside, so bees have more foraging days and are able, on average, to produce more honey.

Where to put your hive

Even small gardens can house a hive, but what applies when considering the placing of a hive in the open countryside is even more crucial when establishing an apiary in town. Some important points to consider are:

1. A small or city garden may have many neighbours who will have opinions on your plans to keep bees. You should talk to them initially if your bee flight paths will pass over their gardens, especially if they have children.

2. Avoid sites that may be too hot. Bees will cease to fly out of the hive above 37°C, so do not choose sites that get full sun during the hottest part of the day. Consider using shading materials or planting to give the bees the benefit of light shade between 11 am and 2 pm.

3. Go upwards. If you have only a small garden, and maybe small children too, consider siting your hives on a flat roof or balcony, so long as it can take the weight of a fully laden hive (c.100lbs) and your weight too.

4. Make extra sure you are insured against theft and for public liability, especially if you are selling honey. Membership of the BBKA includes public liability insurance, which is essential to protect yourself from legal action should a passer-by be stung by one of your bees. Additionally, it is important to consider that your hives may be open to theft or vandalism, although of course this can happen anywhere.

Think carefully about siting your hives and consider the reaction of other people to them if they are on open ground, or even in your own garden.

Wherever they are, it is important that the hives do not border a public highway. Bees need to be able to fly in and out of the hive without obstruction. A barrier placed a couple of metres away from the hive will cause the bees to fly upwards when exiting and so will keep them well above ground level. A 1.5m to 2m tall hedge or screen will work well, and may help to site hives in a small garden.

Face the hives south-east, so that the entrance faces the morning sun – but avoid putting the hive anywhere that a strong prevailing wind will blow into the hive or across the entrance as both will limit the time bees can fly out and forage.

Equally, avoid still air when choosing your site – it will harbour moisture, and damp, cold air in winter will not encourage good health in your bees.

If you are siting hives near to one another, face their entrances in opposite directions to reduce the drift of bees between hives and to limit robbing.

Bees & Beekeeping Explained

Setting Up as a Beekeeper

Siting a hive in a flower border can improve both foraging and pollination of your plants. Face the hive south-east to get the sun into the hive early in the day.

If you do not have the space for your hive, ask local farmers or other landowners as many would be glad of your pollination services. Many farms have small copses, the edges of which will provide a good home to your bees. Sadly, many allotment groups do not allow hives on their sites, but it is always worth asking as this position seems to be changing as the plight of our bees becomes more understood.

Clothing and equipment

There are several pieces of equipment that are essential for the beginner. A toolbox is a handy place to keep your bits and pieces. You will naturally default to using those pieces of kit that suit you best.

A word about hygiene: the possibility of transfer of physical and chemical contaminants, parasites and disease should be at the forefront of the beekeeper's mind when working in the hive. Avoid placing any hive part directly on the ground and do not move between hives without cleaning equipment such as hive tools that will be used in many different hives.

Chemical contamination of the interior of the hive should be avoided at all costs for fear of damaging the brood and contaminating honey. Trading Standards officials will test honey for contaminants such as those that derive from smoke, so always smoke the hive gently and give the bees time to calm down: two or three minutes from lightly smoking an entrance should suffice.

Bees & Beekeeping Explained

Setting Up as a Beekeeper

Bee Suit

A bee suit is essential for the new keeper. Various models are available, from full-body to just jackets. Whatever you decide to use, it is important to have no gaps, and to have loose clothing, as bees will both find gaps and sting through tightly fitting clothes. Dark colours, especially blue, are best avoided as bees seem to be agitated by them.

With use, bee suits will become stained with propolis and pollen, and should be cleaned regularly using a simple detergent that has little or no scent that might cause the bees to react against it.

Gloves

All beekeepers will wear gloves at some point, and most will much of the time. Thin gloves have the advantage of allowing greater dexterity when handling the bees, but these are more likely to be stung through than thicker hide or rubber gloves. Wear what suits you and, again, clean them regularly to avoid transmitting contaminants or disease between hives.

Hive tool

This multi-tool is of real use when opening hives and shifting frames within supers and brood chambers. One end is sharpened for cleaning wax from frames, whilst the other end is hooked for lifting frames.

Choose one that has a hole for attaching a piece of, ideally brightly coloured, string (which helps locate it if it is dropped in long grass).

Smoker

Smoke is really the only tool the beekeeper has for controlling the behaviour of bees when working with the colony. Smoke causes the bees to gorge themselves on honey, reacting to the potential threat of fire – the bees are therefore both distracted and made less likely to sting the keeper.

Smoke varies, though, and natural materials such as dried grass, hessian or dried rotten wood are far preferable to cardboard, which may well carry contaminants and is not recommended.

If smoking the bees does not calm them, it is likely that you need to check if you have lost the queen, or buy yourself a calmer strain of bees.

When using your smoker, always top the smouldering material with fresh, green grass to cool the smoke as it exits.

Clean sheets

A piece of clean cotton can be handy when working in the hive. Many keepers cover portions of the hive with a cloth to keep particular sections of the hive in the dark when working on others – this is especially useful in the brood chamber where you might want to prevent cool air passing over the brood. A crown board can be used in this manner too.

Preparing your honey for sale

Should you be planning to bottle and sell your own honey, which most

---

The beekeeper's toolkit: (clockwise from bottom centre) lighter, dried rotten wood for the smoker, toolbox showing soft brush, queen cage, hive tool, antihistamine; smoker standing on table. A mobile phone should also be carried if working away from home – you never know when you are going to need a hand carrying all that honey!

keepers do, then you must be aware of both food handling legislation and those parts of trading standards legislation which apply to honey.

Various government websites give helpful information, such as the Trading Standards Institute. The BBKA site is also useful.

Honey being sold for public consumption must be produced in a hygienic and safe environment with no risk of contamination, and honey stored once extracted should be kept cool and dark when bottled to prevent any deterioration.

It is said that honey will not spoil, but this is not strictly the case. Honey that is of sufficiently low moisture content – below 20% water – will keep without fermenting so long as it is properly sealed, but honey does alter in composition with time.

The amount of an enzyme, diastase, decreases with time, and therefore the quantity of it present in a sample can be used to detect the approximate age of the honey in question.

The presence of another chemical, HMF (hydroxymethylfurfural), can also be used to tell if the honey has been unduly heated in its production and bottling and may be cause for action by the authorities. HMF is present in tiny amounts in all honey, and increases with any temperature treatment of the honey and the time period involved, so any use of heat when preparing honey for sale should take place as quickly as possible.

Once collected and bottled, honey should be stored and used within one year of production.

Granulation

Because of the nature of honey and its tendency to granulate in storage, the

Straining honey from the extractor into a food grade storage bucket.

extraction of honey from the comb should take place as quickly as possible after the honey is harvested, and ideally on warm days when the honey in the comb will be as fluid as possible.

Different honeys granulate at different rates – those with a high glucose content such as oilseed rape do so quickly. Beekeepers must constantly check for readiness, and be prepared to extract at short notice.

Once honey granulates in the comb, it is messy and difficult to extract, and the comb will be ruined. Ideally, once honey is extracted, the comb should be given back to the same bees to clean and repair, and refill if possible. Having said this, many beekeepers do not keep comb for more than one or two seasons, preferring to melt it down and replace the frames to limit disease and contamination.

Setting Up as a Beekeeper

Extraction

Some thought should be given to how you will extract your honey once it is gathered. The first port of call should be your bee club as it is likely that they have equipment that can be loaned or borrowed. Failing that, a manual honey extractor can be bought for around £100 and filters and jars are available on many beekeeping web sites. The details of using particular types of extractor are not referred to here due to the sheer number of types.

When you are ready to extract, make sure that your equipment is clean and that you are too. A helper who can remove the fine cappings from the honeycomb is handy. The wax capping on the comb is fine and of high quality and should be kept separate from other wax, washed well to clean it and reserved for polish manufacture. Remove it carefully from the comb to limit damage, using a sharp carving knife or heated wax knife.

Labelling

The regulations relating to honey labelling are detailed, but not complex. The BBKA website carries all the relevant information. It should be noted, for instance, that labels must contain not only the information required such as the producer's name, but also that the information should appear in a certain size of character.

Frames of honey in an extractor.

CHAPTER 6

Bee Diseases and Related Problems

Bee health is determined by many factors. Within these, food availability, climatic conditions and inbreeding are those that affect the colonies' ability to thrive and perhaps resist attack by all but the most vigorous diseases.

It has already been noted that honeybees exist in the UK almost exclusively in managed hives. As a result, it is the responsibility of the beekeeping community to maintain their colonies in as good health as possible, not just for their own sake, but for the sake of neighbouring beekeepers and the wider population as a whole, whose food supply depends in large part on the role of the honeybee and its relatives.

Modern bee transport, and indeed the development of the movable frame hive itself, has resulted in bee diseases being transported at much faster rates than previously would have been possible. Indeed, the mass transport of hives across large distances and the transport of breeding stock between continents has led to an unprecedented threat to the world honeybee stock.

This chapter deals with those problems, parasites and diseases that most regularly affect honeybees in the UK. Whilst every attempt has been made to ensure the information given is up to date, every beekeeper should update his or her knowledge regularly.

The BBKA website has regular updates on disease management, as does that of the National Bee Unit, which is managed through FERA, the Government's Food and Environment Research Agency. The NBU's role is to monitor disease outbreaks and threats and to advise beekeepers on these. The inspectorate is especially useful in helping beekeepers monitor disease in their hives and is able to advise on treatments. A free inspection service is available through their website.

Food related problems

So long as food and water are available to the honeybee, a colony will grow and produce brood. Should bees not be able to feed for any reason, the colony will reduce brood production and eventually starve.

It is the responsibility of the beekeeper to ensure that honeybee colonies have enough food at all times of the year, as even during the summer months, if the weather is not good, growing colonies will rapidly fail if food is in short supply.

Dysentery is a problem associated with poor quality food within the hive. It can happen because of honey granulating within the comb or because autumn syrup ferments in the comb. In the former, high glucose

Bee Diseases and Related Problems

A feeder such as this can be used within the hive to provide additional food for the bees in time of food storage.

honey such as that from oilseed rape crystallises, leaving a weak fructose solution that may ferment between the glucose crystals. Eating this watery solution causes the bees to evacuate their rectum, often within the hive in cold conditions, and can cause hive losses in severe winters.

On the other hand, if autumn sugar syrup is fed too late, the bees are unable to ripen the syrup to a sufficiently high concentration before they cluster, with the result that the syrup will ferment and cause similar problems.

Bees leaving the hive in the winter for evacuation flights emerge streaking dysentery down the front of the hive, which can easily be observed. There is little that can be done to cure the problem, and beekeepers can only hope for a dry spring to provide good quality foraging for the growing colony.

Poisoning

Honeybees can suffer from poisoning by chemical treatments given to nearby crops. Recent mass losses in Germany seem to have stemmed from the misapplication of neonicotinoid pesticides to seeds.

A sign of poisoning is the finding of dead bees in front of the hive. Once discovered, a sample of a couple of hundred bees should be sent to the National Bee Unit for diagnosis. Clearly, pesticides should never be used near to colonies.

Parasites of the honeybee

VARROA DESTRUCTOR

This exotic pest has been a recent introduction to the UK, but its spread has devastated colonies of *Apis mellifera* worldwide. No more serious or immediate a threat exists to honeybees. Around the world, many research programmes exist to discover any possible cure.

A parasitic mite, *Varroa* reproduces within the hive, feeding on both larval and adult bees by sucking their haemolymph (bee blood). The mite has been a known pest of the Asian honeybee, *Apis cerana* for a hundred years, but this bee exhibits aggressive behaviour to the mite and kills it once it enters the hive.

Apis mellifera does not cope well with *Varroa*. The bee larva is weakened severely by the action of the mite within the brood cell, and may well have many viruses introduced through the mite's sucking mouthparts. If the larva survives and pupates into an adult bee, it will be weak and may exhibit one of the many bee viruses that *Varroa* transmits, such as Deformed Wing Virus.

The management of *Varroa* follows the principles of integrated pest control, attacking the pest at several different levels in an attempt to limit the possibility of the mite becoming resistant to any one method. Some of the control methods are outlined in brief below. It is necessary for all keepers to practise active *Varroa* management.

Monitoring *Varroa* through the use of mite drop is discussed in detail on the NBU *Varroa* information website.

Bee Diseases and Related Problems

A shallow frame inserted into the brood chamber will attract new drone cells beneath. This can be removed as part of Varroa monitoring, as the mite prefers to lay in drone cells.

The use of mesh floors, monitoring screens and drone removal are all methods by which beekeepers keep a close eye on the presence of this parasite in the hive.

Adult female *Varroa* prefer to lay their eggs in drone brood cells as the drones take longer to develop, providing food for the larval mite for longer. By regularly uncapping, and removing some drone brood if necessary, the beekeeper can monitor mite build-up and treat accordingly.

Encouraging cleaning behaviour: By opening the hive and dusting the bees with icing sugar, grooming behaviour is encouraged. Through grooming, the bees remove many of the mites present on the adult bees which fall to the floor of the hive, through the mesh floor. The amount of mites that have fallen indicate whether or not chemical treatment is required. Breeding programmes worldwide are focused on developing behaviour in bees which either encourages greater grooming (and therefore mite removal) or behaviour that is aggressive towards the mite such as biting, which is common in *Apis cerana*.

Pyrethroids: These basic insecticides have been widely used, and misused, and the UK *Varroa* population is quickly becoming resistant to them. The NBU website will tell you whether mites in your area are becoming resistant. If you are planning to treat with these chemicals, you must ensure that no honey is present in the supers and that the manufacturer's instructions are followed to the letter.

Oxalic acid: A substance which is naturally in honey, and can be used at high concentrations to treat against *Varroa*. Treatment takes place in midwinter when there is very little brood in the hive, so that no *Varroa* can hide within the brood cells and avoid the chemical. Almost all mites in the hive are killed.

Thymol: Has a good reputation and the chemical will penetrate beeswax, so is effective even when mites are in the brood cells. Almost all the mites in the hive are killed.

Essential oils and herbal teas: Some beekeepers have attempted to use

Bee Diseases and Related Problems

Wax moth damage on stored comb – note the silken tunnels and dark faeces.

essential oils such as wintergreen to treat against *Varroa*, while others use camomile and other herb teas in an attempt to improve bee health.

Small hive beetle

An exotic pest not yet considered present in the UK, this small beetle is potentially a huge threat. Endemic to Africa, the beetle has spread worldwide, causing large colony losses. Larvae burrow through comb, eating honey and brood.

Bee mites – *Braula*

Many species of bee mite are widespread across many continents. They can harm the bee if present in large numbers, but generally are considered only a pest. They crowd around the bee's mouth and eat pollen and nectar as she feeds.

Acarine

A small tracheal mite which affects relatively small numbers of hives in the UK. This mite has caused large losses in the past in other countries. No current chemical control is available in the UK.

Wax moth

Both the lesser and greater wax moths cause huge damage to comb, especially when stored if untreated. The large wax moth is present in most of the southern counties of the UK, yet only at the northern extreme of its habitat. Both types of moth are susceptible to cold, and an overnight treatment in the freezer before comb is stored is an effective control. Healthy, strong

Bees & Beekeeping Explained

A hive made ready for winter, with mouse-guard in place.

colonies are more able to withstand attack, so small colonies should be more appropriately housed in nucleus boxes.

Mice and wasps

Mice are pests of hives during the winter and gain entry when the bees have clustered and are not provoked to sting the invader. They use hives to shelter from winter weather and raid them for food, eating honey and wax, and cause huge damage. Mouse-guards can be fitted to prevent their entry during the winter.

Wasps present a potentially lethal threat to hives. Adult wasps will raid colonies and steal both brood and adult bees to feed their own brood during the early and mid summer. Later, wasps will raid hives for honey and may attack in concert, overpowering small colonies or those with wide entrances, which are not easily defended. Care should be taken not to attract wasps to the apiary and to destroy nests nearby.

Diseases

Chalk brood

A stress-related disease caused by the

fungus *Ascosphaera apis* which affects the larva, killing it. Hives that are well ventilated and stress-free seem to be less affected.

American Foul Brood

Colonies affected by American Foul Brood do not usually survive and, indeed, must be destroyed before other colonies nearby are affected. The disease is characterised by sunken, blackened brood cells, with glue-like rotten larvae, and is easily transmitted within and between hives. In the UK, a beekeeper must inform the National Bee Unit if a hive shows characteristics of this disease, and the local bee inspector will advise on treatment, which usually consists of burning the affected hives.

European Foul Brood

Despite the name, this is not as serious as American Foul Brood; this disease may pass through the apiary and go unnoticed. More common in the spring, it is only a serious problem in a small number of areas. The appearance of affected brood cells is similar to that in American Foul Brood. Differentiating the two diseases is not easy for the novice and if either disease is suspected, the National Bee Unit should be contacted immediately.

Nosema

This is a protozoan parasite that is transmitted through faecal matter within the hive, and one that may persist for several seasons. A widespread disease of adult bees, it causes premature ageing and mortality. It can, however, be treated using Fumidil B.

Colony Collapse Disorder

It may well be fitting to include a section on Colony Collapse Disorder (CCD) beneath a note about Nosema, as many hives affected by this curious problem do indeed seem to have Nosema present in the bees left in the hive.

Huge losses in the numbers of hives, especially in the USA, have been attributed to this syndrome in which hives are found to be abandoned with little or no brood. Affected hives are usually left untouched by nearby colonies which might otherwise be expected to rob the remaining honey stores. Tens of thousands of hives have been lost worldwide in the last three or four years, and yet it is still not apparent whether CCD has a single cause.

It cannot be denied that honeybees have suffered from acute environmental stresses, exacerbated when hives are repeatedly moved for the purpose of crop-pollination. Organic methods advocate that hives are not moved unduly to limit stresses on the bees.

It is possible that various environmental factors combined with a series of long, cool, damp winters have caused losses in the UK, and that these losses may not be related to one disease at all. It may simply be that honeybees are constantly under attack from many sides, and one small environmental push is all that is needed to send them over the edge. As usual, well looked after colonies that are properly fed and treated for disease should survive most attacks.

Much research continues, and the National Bee Unit should be contacted immediately if a hive is found to have been suddenly abandoned.

Chapter 7

Gardening for Bees

Even if you are not a beekeeper, any gardener can grow for the benefit of bees. Imagine how much more foraging would be out there if all gardeners, county councils and farmers planted and grew for bees!

By gardening for bees, the gardener helps not only honeybees of course, but many other insects that benefit from having nectar and pollen-rich varieties of flower available throughout the year.

That last point is key because in many of our rural environments there is no longer the biodiversity to provide foraging for bees that there was even 60 years ago, due to the loss of habitat. The 'green revolution' that began in the Americas and shifted across the world in an attempt to produce more food for man has led to a loss in the foraging species available to bees and other pollinating insects.

It is not necessarily the size of your garden that matters. Many bee species will forage extensively, searching for a wide variety of food plants. Even honeybees will forage extensively outside the times of major honey flow – essentially outside the summer months. Window boxes, tubs and planters filled with bee-friendly plants might appear to be small fry in the greater scheme of things, but every bit helps. Those species that repeat-flower are a great help in all of this. All local authorities are required now to have a biodiversity action plan, so our roundabouts, parks and managed gardens ought to soon show signs of bee-friendly planting.

The lack of available forage for bees in rural situations can be remedied in part by feeding, and this is of course necessary in areas such as heather moorland where, if the bees are not moved onto lowland foraging in the earlier part of the year, there is only a short period of wild foraging available. Feeding may always be necessary for some bees in some situations, but planting for bees can help.

Below I have given a rough guide to bee-friendly plants, but you should remember that not all plants secrete nectar in all years, and some plants of course are much more important as pollen producers than others. For example, willow is important as an early season crop that is helpful to the bee when the brood nest is building after the winter's dearth.

Flower shape and complexity is a significant consideration when choosing plants. Honeybees have rather short tongues compared to bumblebees, butterflies and moths. Therefore, complex – i.e. double flowers and those that have very long flower tubes – are not suitable for honeybees as a food source. Simple, open flowers are the easiest for honeybees to access. Think apple, dandelion and borage.

The quantity you plant is important. Aristotle recognised that honeybees are driven to visit one plant species

Gardening for Bees

at a time and the gardener can plan with this in mind. It might be better, for example, to have a drift of one plant species per month or season to add to the foraging that is otherwise available in the locality. A square metre or two of a particular species will benefit the bees more than one individual specimen.

Climbers and small trees can be used to give additional foraging space in the smaller garden. Ivy is a particularly useful forage crop for honeybees, flowering as it does so late in the season, and providing both nectar and pollen in mild years. And, of course, plants can benefit both honeybees and your kitchen garden, with fruits such as apples, raspberries and strawberries being obvious delights.

Greater species diversity leads to tastier honey, but the beekeeper is not able to alter the bees' basic ecology. A colony will collect as much food as possible, and honeybees are very effective at communicating to one another where a food source is. Given the best will in the world, and the most attractive garden of bee-friendly species to feed from, there is nothing that will stop bees heading off to a field of nearby crops, where the massive potential for foraging far exceeds that in their immediate vicinity.

Many rural beekeepers will gather oilseed rape honey in May and June – a simple, vegetal honey that has a more delicate taste than some others. At that time of year, although there are many other wild flowers in the hedgerows, bees will focus on the rape. For the bee, it's a no-brainer. At other times of year, especially early and later in the season, we are more likely to see honeybees in our gardens. It is at these times of year that the gardener can be of use to the bee.

Should you be interested in helping specific bee species other than the honeybee, then the Bumblebee Conservation Trust offers excellent advice. Their web site, www.bumblebeeconservation.org.uk gives a bumblebee specific range of plants that can help the gardener choose what to plant in addition to my list for honeybees. Some of our bee species prefer to winter in hollow stems such as those of hemlock and reeds. Tidy, urban gardens can often have little habitat for these bees, so why not consider a mason-bee nest of the sort available from henandhammock.co.uk which offer eco-friendly garden supplies for all. And, remember, even if you only have a windowsill, you can still help.

Fields of oilseed rape in Lincolnshire.

Plant list by month

Early spring flowers can provide useful pollen for brood feeding – note the orange pollen in the baskets on the legs of this worker.

January and February
Hellebores such as the Christmas rose
Snowdrops
Winter aconites

March
Willow
Early spring bulbs
Japanese quince
Coltsfoot
Lesser celandine
Willow

April
Spring bulbs
Forget-me-not
Alyssum species
Blackcurrants and other *Ribes* species such as flowering currants
Gooseberry
Hazel
Sycamore
Red deadnettles

May
Blackthorn – sloes and all plums

Dandelions

Gardening for Bees

Crab apples
All apples and pears
Dandelions
Brambles
Hawthorn
Wallflowers
Catmint
Broad beans
Cotoneaster
Gorse
Marsh marigolds

June

All apples and pears
All late flowering plums
and cherries
Borage
Brambles

Rosemary

Mint
Rosemary
Runner beans
Sage
Thyme

July

Almond
Dogwood
Californian poppy

Lavender
Lime
Hogweed
Poppies
Thistles
Valerian
Meadowsweet
Chickweed
Field scabious
Loganberries
Borage
Sweet bay
Geums

A bumble bee on a bramble.

Raspberries
Tayberries
Holly
Horse chestnuts
Honesty
Campanulas
Hyssop
Marjoram
Marrow

Californian poppies

Bees & Beekeeping Explained

Wild strawberry flower

Candytuft
Verbena
Roses – simple flowers

Red valerian

August

Love in a Mist
Mignonette
Poached egg plant
Hollyhocks
Heather
Clover
Sunflowers
Globe thistles
Sea holly
Scabious
Teasles
Salvias – all sages are very good
Orange ball buddleia
Strawberry tree – *Arbutus unedo*

Ivy flowers

September

Dahlias – simple single flowers
Eryngiums

Gardening for Bees

Globe thistles
Cardoons
Sedums
Rosebay willow herb
Japanese anemones
Michaelmas daisies
Snowberry
Mallows
Bindweed

October, November and December

Ivy
Mahonias
Clematis *cirrhosa* – throughout the winter
Christmas box – *Sarcococca* species

Mahonia

BEES & BEEKEEPING EXPLAINED

GLOSSARY

City rooftop hives

Anther	The male organ in a flower that produces the pollen necessary for the insemination of other flowers in the same plant species and, therefore, the production of viable seed.
Beeswax	Beeswax is a substance produced from wax glands that are present on the ventral surface of the worker bee's abdomen. Beeswax is produced in small scales that are chewed by honeybees and is thereby formed into honeycomb.
Brood	The term given to the young developing bees from the stage of the egg to the time that the larval bee pupates and emerges into the adult insect.
Brood chamber	That part of a beehive that contains the queen, eggs and young developing bees.
Brood food	The secretions made by worker honeybees, which are used to feed worker and drone bees.
Colony	A colony of bees, be it in a hive or hole in a tree, consists of one queen and her female and male progeny. A fully formed colony in midsummer numbers approximately 50,000 bees.
Drone bee	A male bee. The product of an unfertilised egg laid by a queen bee.
Enzyme	An enzyme is a complex protein that acts on substances to alter the rate of a chemical reaction. Honeybees secrete an enzyme called invertase that converts the sugars in nectar to honey.
Extra-floral nectarines	The name given to nectar-producing organs that sometimes occur outside the flowers of a plant, on the stem of field beans for example.
Fat body	Cells within the abdomen store fat, glycogen (carbohydrate food store), and sometimes reserve

Glossary

protein. This body grows in size in the autumn and allows the worker bee to survive the winter.

Foundation The flat sheets of beeswax that beekeepers give to bees to make into honeycomb.

Frame Beekeepers give bees sheets of wax suspended in frames on which to build honeycomb. Frames enable honeycomb to be lifted in and out of the hive.

Hive A hive is the wooden construction in which beekeepers keep bees. Each hive usually contains just one colony of bees.

Honey Honey is produced by the concentration of nectar by honeybees, and the conversions of certain sugars within the nectar to ones which can be concentrated and stored as food.

Honeycomb The wax produced by bees that is manipulated into hexagonal tubes in which the bees both rear young (brood) and store food.

Honeydew Honeydew is a derivative of plant sap. Aphids tap the sap-rich stems of many plant species and exude the sugar-rich sap, which honeybees collect and store in the same way that they do nectar.

Nectar A derivative of plant sap that is rich in sugars, and which is produced to attract flying insects such as honeybees that, in return, pollinate the flower.

Nectary The organ within the flower of a plant which produces nectar.

Nucleus of bees A nucleus of bees is a small, usually five-framed, colony which is produced by a beekeeper for sale or for increasing their stock. A small box is used to avoid stressing the growing colony.

Ovary The female part of a flower which must be inseminated by pollen from the male anther of the same or another flower in order to produce viable seed.

Pheromone A chemical produced by a living organism which has an effect outside the body, usually on other individuals of the same species.

Pollen Pollen is produced by the male parts of a flower, the anthers,

	and is rich in protein. It is used by bees as a food source, and is transferred by bees from flower to flower of the same species, thereby pollinating the female parts of the plant – the ovaries.	Queenright	The term used to describe a colony which has a queen.
		Royal jelly	A substance produced by worker bees to feed the developing queen bees. It is much more highly nutritious than brood food and is thought to account for the size and longevity of the queen bee.
Propolis	A resin-like substance which worker honeybees collect from various species of shrub and tree. Flexible when warm in the hive interior, it is used to fill gaps and to coat the inner surface of the hive. It has anti-microbial properties.		
		Spermatheca	The organ within the queen in which she is able to store sperm from her nuptual flight, which she does until she either runs out or dies.
Protozoa	Protozoa are simple, single-celled animals that may cause disease.	Super	That part of the hive above the brood chamber that is used to store honey, and which is usually separated from the brood chamber by a queen excluder to prevent brood rearing and honey production occurring in the same part of the hive.
Queen bee	A queen bee develops from a fertilised, female egg which is fed special food called royal jelly by worker bees. Usually, there is only one queen bee in each colony or hive.		
		Worker bee	A female bee that usually does not lay eggs. The majority of the bees in a colony or hive are workers.
Queen excluder	A perforated metal sheet that is used above the brood chamber, and which only worker honeybees can pass. It is used to prevent the queen laying eggs amongst the honey-storage part of the hive, the *super*.		

Useful Addresses

Useful Addresses

British Beekeepers' Association,
National Beekeeping Centre
National Agricultural Centre
Stoneleigh
Kenilworth
CV8 2LG
www.britishbee.org.uk

Natural Beekeeping Trust
www.naturalbeekeepingtrust.org

The Co-operative Group

The Co-op are currently supporting urban beekeepers, in particular, with their Plan Bee initiative to encourage their members and customers to help the honeybee by becoming bee-friendly gardeners and offering bee-keeping courses. Following a trial in Manchester, they are rolling out training programmes across the U.K. starting this year. For further details, visit their website:
www.co-operative.coop/ethicsinaction/takeaction/planbee/news/latest-news/

Manufacturers of beekeeping equipment:

National Bee Supplies
Merrivale Road
Exeter Road Industrial Estate
Okehampton
EX20 1UD
Tel. 01837 54084
www.beekeeping.co.uk

E.H. Thorne Ltd
Beehive Works
Wragby
Market Rasen
LN8 5LA
Tel. 01673 858555
www.thorne.co.uk

Bee Basic Ltd
5 Hillcrest Avenue
Pinner
HA5 1AJ
Tel. 020 88663864
www.beebasic.co.uk

BB Wear
1 Glyn Way
Threemilestone
Truro
TR3 6DT
Tel. 01872 273693
Email: Mike.bbl@btinternet.co.uk

Fragile Planet Ltd
Unit 14 Radfords Field
Oswestry
SY10 8RA
Tel. 01691 672869
www.fragile-planet.co.uk

An Italian yellow bee on a winter aconite.

Index

acarine 51
American Foul Brood 53

bee mites 51
bee space 37, 38
British Beekeepers Association 38, 44, 45, 46
bumblebee 6–7
Bumblebee Conservation Trust 55

Chalk Brood 52
clothing 43
colony, established 29, 30, 31, 32; inside the 7–12; new 28, 30, 31, 32
Colony Collapse Disorder 53
communication 13–15
contamination 19, 41

disease management 29, 30, 32, 41, 46, 52
drones 7, 8, 12
dysentery 46–47

equipment 41–43
essential oils 50
European Foul Brood 53

feeding 27, 29, 30, 31, 32, 46
flowers, choosing 54–55; list of 55–59

grooming 50

herbs 51
hive, construction 36, 37, 38; siting 39–40; temperature 12, 15, 27
hive tool 43
honey, adulteration 19; constituents 19; extraction 45; granulation 44; labelling 45; making 15–19; selling 43, 44, 45
honeydew 16

insurance 39

mice 52

Nasonov gland 15
National Bee Unit 46, 48, 50, 53
nectar 15–17
Nosema 53
nucleus 28–29, 61

oxalic acid 32, 50

parasites 48–52
pheromones 8, 9, 12, 15, 25, 33
plant sap 16
poisoning 48
pollen 21
propolis 21
pyrethroids 50

queen bees 7, 8, 25–26; mating flight 8; replacement of 8, 25, 33

royal jelly 8, 11, 21, 62

small hive beetle 51
smoker 43
spermatheca 8, 62
stings 33
super frame 21, 29, 30, 31, 62
swarming 8, 22–27, 29, 32

thymol 50
training 38

urban beekeeping 16, 38–39, 60

Varroa Destructor 32, 48–51
venom 11

waggle dance 14–15
wasps 52
water 10, 11, 15
wax 11, 12, 20, 29, 45, 60
wax moth 51
winter survival 7, 15, 27, 30
worker bees 7, 8, 9–12; communication by 14